Zoubir Layouni

New problem results: Consecutive Block Minimization

Zoubir Layouni

New problem results: Consecutive Block Minimization

ScienciaScripts

Imprint
Any brand names and product names mentioned in this book are subject to trademark, brand or patent protection and are trademarks or registered trademarks of their respective holders. The use of brand names, product names, common names, trade names, product descriptions etc. even without a particular marking in this work is in no way to be construed to mean that such names may be regarded as unrestricted in respect of trademark and brand protection legislation and could thus be used by anyone.

Cover image: www.ingimage.com

This book is a translation from the original published under ISBN 978-620-6-73289-1.

Publisher:
Sciencia Scripts
is a trademark of
Dodo Books Indian Ocean Ltd. and OmniScriptum S.R.L publishing group

120 High Road, East Finchley, London, N2 9ED, United Kingdom
Str. Armeneasca 28/1, office 1, Chisinau MD-2012, Republic of Moldova, Europe
Managing Directors: Ieva Konstantinova, Victoria Ursu
info@omniscriptum.com

Printed at: see last page
ISBN: 978-620-8-99217-0

Contents

Introduction

Throughout this thesis, we will be concerned with a simple structure: a binary matrix. Let .4 be an m X -m-binary *matrix*. We will be interested in a special property, known as the "1-consecuity property", or PC1 for short. Consider any line ¿ in .4. A consecutive block is a sequence of 1's located consecutively (we'll give a formal definition in Chapter 1). The problem we are studying here (known in the English-language literature as "Consecutive Block Minimization" or CBM for short) consists of finding a permutation of the columns of .4 so that the number of consecutive blocks in the induced matrix is a minimum.

After posing the CBM problem in Chapter 1, we recall that it is NP-complete for general instances, then we present the applications which concern it, the variants and a state of the art. All this is based on a rich bibliography.

The CBM problem is well known to be NP-complete for a general instance. One of the usual approaches to an NP-complete problem is to look for particular instances for which the problem becomes "easy", or conversely, to characterise particular instances for which the problem remains "hard". Our first contribution, in Chapter 2, is to prove that CBM is NP-complete even when the binary matrix has only two 1's per row, by polynomially transforming the maximum weight Hamiltonian chame problem to CBM restricted to the instances in question.

The question of the G-approximation of CBM was open (to the best of our knowledge). A second contribution consisted in solving this question: is CBM approximable with guarantee? The answer was found in the form of a polynomial heuristic that constructs permutations resulting in a number of consecutive blocks that deviates by no more than 50% from the optimum. This work has been published.

Chapter 1

Position of the problem, complexity, applications and variants

1.1 Introduction

The problem we are going to present in this thesis consists in defining a problem called CBM, from the English term "Consecutive block minimisation". It consists of obtaining an acceptable polynomial solution by minimising the number of consecutive blocks of 1.

The CBM problem is well known in the scientific community [1,3,8,9,10,11,12], and a very extensive bibliography is provided as a reference. It is an NP-complete problem [13], however, it has been shown that it is also NP-complete even if we restrict ourselves to binary matrices with two 1s per row. [14]

Given a binary matrix -4, formally a consecutive block is an arbitrary sequence of coefficients . $A_{ip}, A_{ip+1}, \ldots, A_{iq}$ of the line ¿ satisfying

(i) $A_{ij} = 1, p \leq j \leq q,$

(ii) or o= I or $A_{i,p-1} = 0,$

(iii)or ç= nou bien $A_{i,q+1} = 0.$

A binary matrix is said to have intervals if each of its rows contains only one consecutive block. For example, the following matrix :

Has 5 consecutive blocks and is therefore nσtot an interval matrix. On the other hand, the

$$A = \begin{pmatrix} 0 & 0 & 1 & 1 \\ 1 & 0 & 1 & 1 \\ 1 & 1 & 1 & 0 \\ 0 & 1 & 0 & 0 \end{pmatrix} \quad (1.1)$$

Matrix

$$B = \begin{pmatrix} 0 & 0 & 1 & 1 \\ 0 & 1 & 1 & 1 \\ 1 & 1 & 1 & 0 \\ 1 & 0 & 0 & 0 \end{pmatrix} \qquad (1.2)$$

is an interval matrix.

Given a binary matrix A, let $A(\pi)$ be the matrix induced by a permutation π of the columns of A. The matrix A is said to have the 1-consecutive property (PCI for short) if there exists a permutation π such that the binary matrix $A(\pi)$ is an interval matrix. The term $n(A(\pi))$ denotes the number of coefficients $a_{i\pi(j)}$ such that $a_{i\pi(j)} = 1$ and $(a_{i\pi(j+1)} = 0 \text{ ou } j = n)$. Let us observe that the number of consecutive blocks is in bijective correspondence with the number $n\ (A(\pi))$. As an example, the matrix A given in (1.1) is an interval matrix since the permutation (2 1 3 4) gives an interval matrix, the matrix B in (1 2). A first problem arises.

Recognise the property of consecutive 1's (PC1) :

Interval graphs are closely related to matrices with the consecutive 1 property. A matrix (0,1) has the property of consecutive 1's if its columns can be permuted so that the 1's in each row appear at consecutive positions.

The characteristic of interval graphs and matrices with this property has been studied by many researchers and a very rich bibliography has been compiled in the appendix, particularly Booth and Lueker [3] who have developed a linear complexity algorithm which recognises the property of consecutive 1's.

Instance: Two integers *m* and *n* and a binary mxn-matrix *A* ;

Question: Is there a permutation π of the set {1,...,n} such that $A(\pi)$ is an interval matrix?

This problem has been well solved by Booth and Lueker [3] using a structure called PQ-trees. The proposed algorithm is linear in O(*n*+ f), where *f* is the number of 1's in *A*.

The interest of interval matrices, and therefore their recognition, lies in their applications (see section 1.3), and the fact that they are graph matrices, and therefore totally unimodular. The interest of total unimodularity for a binary matrix lies in the fact that, when faced with a combinatorial optimisation problem, the polyhedron of solutions has all its vertices whole; consequently, the method of ellipsoids, for example, enables it to be solved efficiently.

If, now, the matrix is not recognised as having the PC1 property, this means that any permutation of its columns results in a number of consecutive blocks greater than the number of rows *m*. Nevertheless, we often want the permutation that deviates least from the PC1 property. This justifies the definition of the problem that concerns us:

Consecutive Block Minimization (CBM) ;

Instance: two integers *m* and *n*, and a binary *mxn-matrix A* ;

Question: Is there a permutation π^* *of* the set {1,...,n} such that $n(A_{\pi^*}) < K$?

In fact, the CBM problem we have just defined is the decision problem associated with the optimisation problem having the same generic instance as CBM, but whose question is: "Does there exist a permutation π^* *of* the set {1,...,n} such that $n(A_{\pi^*}) = \min_{\pi} n(A_{\pi})?$ "?

Kou [13] showed that CBM is NP-complete for a general instance. One of the usual approaches, concerning an NP-complete problem, is to look for particular instances for which the problem becomes "easy", or conversely, there are particular instances for which the problem remains "hard". However, the question of the ε-approximation of CBM remains.

1.2 Complexity

Complexity theory aims to determine whether the answer to a problem can be given very efficiently. To do this, it establishes hierarchies of difficulty between algorithmic problems, the levels of which are called *"complexity classes"*.

We will say that a decision or optimisation problem that takes a graph as input can be solved in linear time and space if there is an algorithm that can solve this problem in time and space. This linear time-space algorithm is the best we can hope for to deal with such a problem.

The problem of the travelling salesman, who seeks, in a graph whose edges are labelled by costs, to find a cycle, of minimum cost, passing once through each vertex, can be stated as a decision problem as follows: *Is there a cycle passing once through each vertex such that any other cycle passing through all the vertices has a higher cost?* The equivalence of these two problems assumes that the proof of existence is based on a constructive argument, i.e., for example in the case of the travelling salesman, actually provides a cycle of minimum cost in the case where it has been shown that a cycle of minimum cost exists. The problem of the existence of a minimum cost cycle is equivalent to the problem of the travelling salesman, in the sense that if we know how to solve one efficiently, we also know how to solve the other efficiently.

An optimal solution S of a minimisation problem can be approximated in polynomial time and space denoted *a - approximated* if there exists a polynomial algorithm in time and space returning a solution $\bar{S}$ to this same problem such as $\bar{S} \leq \alpha S$; the algorithm in question is said to be approximating and a is its ratio.

In the same way, an optimal solution S of a minimisation problem can be approximated to within a if there exists an algorithm returning a solution $\bar{S}$ such that $\bar{S} \leq S + \alpha$ . . For more details on the notion of approximation in combinatorial optimisation, see the results proposed by Hochbaum. [16].

1.3 Applications

The CBM problem is important and has applications in many fields, for example in biology in general, where applications of the physical mapping of DNA have been introduced. Such problems also find applications in physical mapping with hybrid data for genome sequencing, in the planning of radiotherapy sessions, and we have also found very interesting applications in archaeology, database systems, the scheduling problem, traffic control, memory management and the organisation of records.

The focus is more on applications in which it is necessary to arrange a set of objects linearly, with the restriction that the objects in the subset must be consecutive. The most common applications aim to establish when a chromosome gene is linearly sequenced.

Other applications are based on graph theory, where we want to know or check whether a graph is an interval graph or not. This problem can also be used in file organisation systems to arrange records without duplication (without redundancy) on a medium in such a way that the response to each request (or query) can be retrieved as a set of consecutive records.

We have also found more important applications, such as establishing the physical map of a DNA segment. These problems are formalised in the form of linear sequencing. This is formally described by more or less complicated structures known as PQ and PQR trees presented by Booth and Lueker[3].

As with all major projects, the first stages of the Human Genome Project aim to produce a physical map showing the relative locations of *sequence-tagged sites* (STSs*)*. A sequence-tagged site is basically a DNA fragment that is known to occur only once in the genome. Several hundred thousand STS can be identified. Once we have succeeded in sequencing a long enough fragment, it will be easy to position it, provided it contains at least one STS.

This is summarised by the results obtained in a matrix M, comprising as many rows (clones) and as many columns (probes); the coefficient $M_{ij}$is 1 or 0 depending on whether or not probe *j* reacted with clone *i*.

A certain class of graph has been studied and algorithms for particular classes have appeared in the literature, in particular the recognition of interval graphs. This property has various applications and was mentioned by the archaeologist Petrie.

1.4 Variants and related problems

Many applications of this problem have been published. We will begin by presenting this problem for particular matrices, as well as variants of interest to our study. The reader is invited to consult the bibliography in the appendix for further research into variants in this field. We will see that the problem of this type can be approximated for matrices in which there are not two identical columns and can also be approximated for general matrices, we are also interested in the case of symmetric matrices.

This problem has applications in the physical tracing of DNA, and there are solutions in polynomial time that solve this type of problem. A variant of this problem [2] is presented in the form of an NP-hard problem.

The problem of decomposing an integer matrix into a set of binary matrices with the property of consecutive ones has been studied, and is clearly used in radiotherapy to obtain an intensity matrix [5]. Applications can be drawn from this, for example in transport. In many cases, the authors are particularly interested in minimising the number of matrices used in the decomposition (cardinality), and also in minimising the sum of the coefficients in the decomposition (decomposition time).

The property of consecutive ones has also been widely studied. Kendal [6] was one of the first to mention this property; heuristic methods are proposed, and before him Fulkerson and Gross [7] presented the first polynomial solution. In 1972 Tucker [12] presented a characteristic of the problem based on a configuration of matrices.

It is proved in [11] that the problem remains NP-hard for matrices (2,3) with at most two 1s in each column and at most three 1s in each row, and for matrices (3,2) with at most three 1s in each column and at most two 1s in each row, which also solves a problem posed by M.Tagui and Y.Ganjali [4]. It is also proved that the problem is 0.8-approximable for matrices (2,3) and 0.5- approximable for matrices(3,2) in which each column contains at most two 1s .

A first special case concerning the property of consecutive ones is shown. This is the case of a binary matrix A in which a positive integer k is given.

The question is: Is there a sub-matrix B of A with k columns of A having the property of consecutive 1s? A first sequential algorithm in polynomial time is given by Fulkerson and Gross. [7].

However, the problem of consecutive 1s concerning a sub-matrix is NP-complete. The proof of NP-completeness for this type of problem is given in [4] which shows the NP-completeness of a problem slightly different from the original problem as defined below.

Given the problem of a matrix of consecutive 1s whose instance is a matrix $A(m, n)$ and a positive integer k , is there a sub-matrix $B(m, k)$ of A such that this matrix has the property of consecutive 1s ? The columns of B can be permuted so that on each row, all the 1s are consecutive.

This problem was posed by Booth [3] and defined as a transformation of the Hamiltonian path problem. We will discuss this problem later.

The principle of the CBM problem is to find a permutation of the columns of a binary matrix A that will give a minimal number of coefficients. This problem is defined as follows:

Given a binary matrix of the form $A(m \times n)$. The question is. Is there a permutation such that the number of coefficients of this matrix under certain conditions is equal to a number k? Its formulation is as follows:

Given a binary matrix A, let $A_\pi$be the matrix induced by a permutation π of the columns of A. The term $n(A_\pi)$ indicates the number of coefficients $a_{i\pi(j)}$ such that $a_{i\pi(j)} = 1$ and $(a_{i\pi(j+1)} = 0 \text{ ou } j = n)$. The CBM* problem involves finding a permutation π^* such that $n(A_{\pi^*}) = \min_\pi n(A_\pi)$.

1.5 State of the art

The aim of this state of the art is to establish the results obtained in different scientific achievements dealing with the field, particularly the complexity of the problem.

By definition, a matrix whose coefficients are of the order (0,1) has the property of consecutive 1s for its columns (respectively each row) if its rows (respectively each column) can be permuted so that the 1s of each column (respectively rows) appear consecutively.

Furthermore, the number of rows in this matrix (i.e. the number of maximum cliques) is less than n with equality if and only if the graph has no edges.

Interval graphs as defined in graph theory (These are intersection graphs of a set of intervals of the real line. Each vertex of the interval graph represents an interval of the set and an edge connects two vertices when the two corresponding intervals intersect) are closely related to matrices possessing the property of consecutive 1s. A matrix (0,1) has this property if its columns can be permuted so that the 1's in each row appear at consecutive positions.

The characteristics of interval graphs and matrices with this property have been studied by many researchers. In particular, Booth and Lueker [3] have developed a linear complexity algorithm that recognises this property. They were the first to give a linear time-space algorithm for recognising interval graphs. Their algorithm, based on a complex data structure called a PQ-tree, identifies the property of consecutive 1s. Other simple algorithms based on the breadth-first approach (formerly known as BFS) also exist [15].

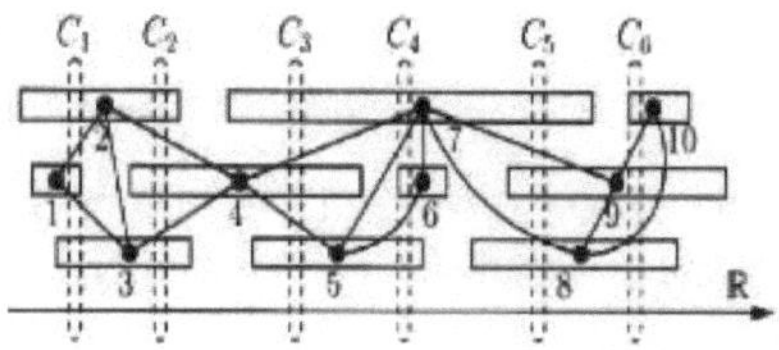

	1	2	3	4	5	6	7	8	9	10
C_1	1	1	1							
C_2		1	1	1						
C_3				1	1		1			
C_4					1	1	1			
C_5							1	1	1	
C_6								1	1	1

Figure 1: Interval graph and its maximum-vs-summary cliques matrix

Before discussing these results, we propose a series of problems that have already been raised in the literature, which we present according to a certain reading logic. First, Booth and Lueker [3] pose the problem as follows. To recognise an interval graph is to recognise the property of consecutive 1s. This means the existence of a polynomial solution which is therefore recognisable in linear time and space. Two important points are therefore known. Determine whether the graph forms an interval graph, and then whether the problem of consecutive 1s for this type of graph is also NP-complete.

Given a matrix $M(m \text{ x } n)$, deciding whether there exists a sub-matrix $(m \text{ x } k)$ with the property of consecutive 1s is NP-complete. This is true if we do a simple reduction of the Hamiltonian path. We have the graph $G(V,E)$, the incidence matrix M admits a Hamiltonian path if and only if there exists $n \text{ x } (n\ \text{-}1)$ sub-matrices M' with the 1s consecutive property. The incidence matrix M for G is dimension $(\text{nx } m)$ where M can have a row for each vertex and a column for each edge and a 1 on the $j'^{\,ith}$element if i is incident on j . Therefore, if there are now $n \text{ X } (n\ \text{-}1)$ sub-matrices with the property of consecutive 1s then there is a permutation P such that PM' also has this property.

There are several cases in which the problem can be solved in polynomial time and space. The main question posed by the authors is when is the graph said to be an interval graph?

A graph is an interval graph if each vertex $v \in V$ is associated with an open interval $I_v$of the real axis such that the pair $u, v \in V$ of vertices, $u, v \in E$ if and only if $I_u \cap I_v \neq \phi$.

A first algorithm by Booth and Lueker uses a complicated data structure called "PQ trees". Its role is to store all the clique chames of the next vertex at each iteration. An interval graph is also a comparability graph, so finding a chame involves calculating a transitive orientation of the complementary graph, which in turn is linked to the problem of modular decomposition. Calculating a modular decomposition of a graph involves substituting a vertex for each module until we obtain a prime graph (with no modules).

This idea does not just mean looking for a permutation P that rearranges a matrix. In the book by Garey and Johnson [27], the authors also propose to construct a permutation form of this matrix column by column by imposing a necessary vector product on the columns of the configuration. Necessary and sufficient conditions are known, however the computational methods are more or less efficient, in addition to the specific representations in terms of intervals that can be easily obtained. A number of important results are presented, the first of which particularly concerns graphs with rigid circuits.

Many improvements have been proposed to the algorithm, however *the* PQ data structure employed is complex and difficult to implement, but when restricted to certain classes of graphs, the problem is easy to solve.

In a graph G, the maximum cliques can be ordered linearly such that for any vertex $v \in V$, the maximum cliques containing v appear consecutively in this order. The maximum cliques -vs- vertices incidence matrix of the graph G has the property of consecutive 1s for the columns, and the number of rows in this matrix (number of cliques) is less than n.

Finally, it should be noted that the recognition of interval graphs in linear time based on the PQ structure makes it possible to identify the property of consecutive 1s. However, any interval graph G that can be represented by an open or closed interval can also be obtained in linear time and space.

Now letA $= (a_{ij})$ be the maximum cliques-vs-sums incidence matrix of the graph. We can consider that the 1s in each column of A appear consecutively. This matrix can be obtained from the interval graph. Sub-classes of intervals are presented and are recognisable in linear time and space.

In the work of R.W.Lang et al [17]. It is shown that for a cubic graph with circumference > 5 admits a Hamiltonian circuit if and only if the matrix $A + I$ can be permuted in the row sense such that each column has at most 2 consecutive blocks of 1s and the block can be viewed in a circular manner where the first row and the last are adjacent rows. It has been proved that for a certain $k \geq 2$ the problem of k consecutive blocks remains NP-complete even if we restrict ourselves to symmetric matrices or matrices with at most 3 consecutive blocks of 1s per row. This result shows that the cubic graph G admits a Hamiltonian circuit if and only if there exists a permutation of the rows for $B = A + I$ whose result is a matrix that has at most 2 consecutive blocks of 1s per column.

The complexity result shows that the Hamiltonian path problem remains NP-complete when restricted to planar related triple graphs with no less than 5 edges.

In the article by Habib et al [15], the authors show some linear recognition algorithms based on a lexicographic search method called (exBFS). These algorithms avoid the use of PQ trees and modular decomposition, particularly for classes of triangulated graphs, interval graphs and convex graphs. exBFS is the first triangulated graph recognition algorithm. It begins by constructing an elimination order, followed by verification (marking test). In the same context, an $O(m + n)$ algorithm based on exBFS is presented for constructing a clique tree that simplifies the verification procedure.

In the family of NP-hard problems, the problem of the consecutive 1s property is raised for different binary matrices $A(0,1)$. The aim is to find a very large number of columns of A to form a sub-matrix with the 1s consecutive property. We will see how the consecutive 1s problem remains NP-hard for several types of these matrices. This type of problem also arises for matrices with specific variants such as the consecutive 1s problem for subsets with some k where we want to know if there exists a sub-matrix B of A with k columns characterised by the consecutive ones property.

Extensive studies have been established for sub-matrices with the property of consecutive 1s, this problem remains NP-hard for several types of matrices such as (2,3) matrices with at most two 1s in each column and at most three 1s in each row, (3,2) matrices with at most three 1s in each column and at most two 1s in each row, proposals for approximation algorithms have been established and proved for some matrices containing at most two 1s in each column. An extension is made for symmetric matrices and important results are presented. The problem of k consecutive blocks is NP-complete for a certain $k \geq 2$

The consecutive 1s problem for sub-matrices is formulated in the following general way: Given a binary matrix A (0,1), find a very large number of columns of A that form a sub-matrix with the property of consecutive 1s. Let be the form (a, b) for binary matrices in which there is at most a (1s) in each column and at most b (1s) in each row. It has been shown that this problem remains NP-hard for matrices (2, 3) and (3, 2). This solved a problem posed by Tagui and Ganjali [4] who devoted a paper to proving that the consecutive 1s problem for sub-matrices is 0 .8-approximable for matrices (2, 3) in which there 18 are no identical columns and 0.5-approximable for matrices $(3, \infty)$, they also proved that the problem is 0.5-approximable for matrices (2, 3) however there exists a certain $\varepsilon \succ 0$ such that the approximation of the consecutive 1s problem for sub-matrices $(\infty.2)$ without the factor $\eta\varepsilon$ (where n is the number of columns of the matrix) is NP-hard.

In the family of NP-complete problems, the version of this problem is one of the first forms of the NP-complete problem posed in the book by Garey and Johnson [27]. However, the proof of NP-completeness is poorly referenced. Recently Tagui and Ganjali [4] gave in their paper the above proof. They also proved that the problem is NP-complete even for sub-matrices in which there are at most two 1s in each column and at most four 1s in each row, on the other hand. The consecutive 1s problem can be solved in polynomial time for binary matrices with two 1s in each row and each column. However, their work raised the question of whether or not the problem remains NP-complete for (3,2) matrices with at most two 1s in each row and at most three 1s in each column and for (2,3) matrices with at most three 1s in each row and at most two 1s in each column.

The problem remains NP-Complete if we refer to the work of Jinsong Tan and Louxin Zhang [22]. They prove NP-Completeness by formulating two interesting NP-Complete problems. The first applies to cubic graphs by considering a Caterpillar tree T and define a special spanning tree as an NP-Complete problem to prove that the problem of consecutive 1s of a sub-matrix is NP-Complete for (2,3) matrices. This process is normally done by a reduction of the Caterpilar spanning tree problem for the graph G . If now T admits a Hamiltonian path, (the path itself being an NP-Complete problem) then the columns corresponding to the edges form ηx $(\eta$ -1) sub-matrices where all 1s are consecutive on each row. The second problem concerning the matrices (3,2) requires a reduction of the induced subgraph. However, the problem remains open since it can be approximated with a constant factor for matrices in which there are at most two 1s on each row.

Studying this problem for matrices with a small number of 1's in the columns or rows is not only interesting in theory but also in practice. This restriction was proposed by Istrail [28], hoping to have a tractable scheme of the problem, but the problem is still difficult to solve for hollow matrices.

Finally, there are 0.8 approximation algorithms in linear time for (2,3) matrices and 0.5 approximation algorithms for (3,2) matrices. These results are very important in the study of physical mapping.

The decomposition approach was used in the work of Hsu and Al [9], using an abstract representation of the interval, and they showed that it would be possible to designate an algorithm for recognising a graph of intervals in O(m + nlog n).

In the article by Meidanis et Al [11], the problem is studied in matrix form and the methods for recognising them are described. Numerous improvements have been made to combinatorial algorithms. Benzer [29] posed the problem from the point of view of graph theory, and proposed a solution using a certain characterisation of interval graphs.

Finally, if we now restrict ourselves to certain classes of graphs, the problem can be solved in linear time and space for two subclasses of interval graphs, proper interval graphs (disjoint cliques) and string graphs.

Heuristic methods have been proposed for the consecutive 1s problem, and Fulkerson and Gross [7] have proposed a polynomial solution. A polynomial-time algorithm for the decomposition problem has been presented and proved to be NP-hard for the unconstrained case. For cardinal decomposition problems, special cases of polynomial-time solution are considered. In their work, Baatar and Al [5] propose a minimisation of the sum of the coefficients in the decomposition (decomposition time) and a minimisation of the number of matrices used in the decomposition (decomposition cardinality).

The matrix has the property of consecutive 1s if the 1s occur consecutively in a single block on each row. The authors propose the integer k which represents the set of indices of all (m x n) matrices composed of consecutive 1s.

The decomposition time problem is NP-hard for the case of the decomposition cardinality problem. Solvable polynomial solutions are considered and heuristics are proposed. The study considers two types of objective functions that can be used to evaluate the decompositions of the matrix of consecutive 1s.

In the case $k' = k$, minimising the decomposition time can be obtained in linear time. Choices *of* k' with important applications and algorithms for minimising the decomposition time are discussed, minimising the decomposition cardinality is defined as an NP-hard problem.

A linear algorithm for the unconstrained case is presented in the work of Baatar et Al [5] to calculate the decomposition time. All the matrices take into account the decomposition. Note that the number of matrices is exponential with respect to the number of rows.

When the decomposition time problem is solved in linear time, the decomposition cardinality problem becomes NP-hard, as proved by Burkard [30] for matrices with at least two rows using reduction from the sum of subsets, so the cardinal decomposition problem is strongly NP-hard even for matrices with a single row.

In another context, given a matrix A and a positive integer multiple of a binary matrix, the decomposition cardinality can be solved in polynomial time for the constrained and unconstrained cases. $(k' = k)$

Heuristic solutions for the general case are proposed. We have focused our attention on the problem of recognising binary matrices with the property of consecutive 1s using the PQ trees and PQR trees structure. Meidanis and Munuera [31] proposed a new theory with a new structure that is able not only to record all valid permutations when the matrix has the property of consecutive 1s but also to highlight the possible obstacles when the property is not retained. They reformulated the problem using a collection of sets by developing this new theory. We have seen that to construct PQR trees in linear time, we must first look for valid permutations if they exist for given sets. We record them in the form of a PQR tree. If these permutations do not exist, then the R nodes of the PQR tree indicate the subsets responsible for this failure. These authors show how to construct a linear-time algorithm for PQR trees that is easy to implement and that will replace the schema of other PQs.

Motivated by practical aspects such as the investigations carried out on the genome project and particularly the studies on DNA molecules, Atkins and Middendortf [21] propose in their article a simplified model for physical mapping with probes which tend to occur rarely along the DNA sequence and show that the problem is NP-Complete for hollow matrices. Moreover, the study of physical mapping with chimeric clones is also NP-complete for this type of matrix. Both problems are variants of the consecutive ones problem and give interesting results for several applications.

Interested in a file management problem, Ghosh [32] introduced the concept of organising files, sorted in consecutive order. The aim was to clarify the relationship between the consecutive ones property and interval graphs in order to develop the necessary and sufficient conditions for the existence of the property.

This problem is NP-Complete and it is assumed that it is possible to determine unambiguously when a given record is relevant to a given query, respecting the set of records, such that for each query, the relevant records are saved in consecutive storage. The problem is NP-complete even for the problem with minimum redundancy.

In the article by Meidanis and Munuera [31], a binary phylogeny problem is introduced which allows us to construct a tree that traces the evolutionary history of a group of taxa for which the characteristics are given in the form of a binary matrix. Gusfield [33] presented an O($m + n$) time algorithm for this type of problem with n taxa and m characteristics. They also show that binary phylogeny exists if and only if the inputs admit a PQ tree without the Q nraud, which immediately gives a linear-time algorithm.

Finally, one class of these problems particularly concerns the k - consecutive 1 problem. The question is whether there exists a certain order of the rows of the matrix A such that each column contains at least k consecutive blocks of 1.

Concerning the k consecutive blocks of circular 1s, the idea is to know if there is an order of the rows of A such that each column contains at least k blocks of 1 with the hypothesis that the columns are circular (the first and last are neighbours).

Let's start by defining a binary matrix of order $M(m,n)$ whose coefficients $M(i, j)$ are either 1s or 0s.

In this matrix, a block of consecutive 1s in column j is an interval of consecutive coefficients $M(i,j), M(i+1,j), \ldots\ldots\ldots M(k,j)$ which are all 1s. Note that the number of blocks in column j is the number of coefficients $M(i, j)$ such that $M(i,j)=1$ and . $M(i+1,j)=0$ $(ou\ i=m)$

Each block is called a linear block (the first row and the last row are not adjacent). If we consider the first and last rows to be adjacent, we have circular blocks. The number of circular blocks in column j of the binary matrix $M(m,n)$ is the number of coefficients $M(i,j)$such that $M(i,j)=1$ and $M(i+1,j)=0$ or $j = m$ and . $M(1,j)=0$.

A graph $G(V,E)$ with $V = (v_{(1)}, v_2, v_n)$ is *k-regular* if each vertex v e Vis of degree $d(v) = k$. The circumference of G denoted by $g(G)$ is the length of the minimal circuit (cycle) in G . The adjacency matrix of G denoted by $A(G)$ is a symmetric binary matrix of order (n x n) such that $A(i, j) = 1$ if and only if $(v_{(i)}\ v_{\cdot})$ e E, the unit matrix denoted by I is a binary matrix (nxn)such that $I(i, j) = 0$ if and only if $i = j$.

The results concerning the Hamiltonian path are :

1- A graph G is k -regular with $k \geq 3$ and $g(G) \geq 5$ contains a path Hamiltonian if and only if the rows of the matrix $A(G) +1$ can be permuted to give at least k consecutive circular blocks of 1s in each column.

2- If G is k -regular except for two vertices a and b which are of degree k , then G contains a Hamiltonian path between a and b if and only if the rows of the matrix $A(G) +1$ can be permuted so that each column has at least k consecutive linear blocks.

These two results are achieved by replacing the word "at least" with the word "exactly".

Several variants on the total number of consecutive blocks of 1s are NP-Complete problems. An example is given in the work of Wang and Al [17] for a cubic graph with circumference ≥ 5. . This graph admits a Hamiltonian circuit 24 if and only if the matrix $A +1$ can be permuted in the row sense such that each column has at most 2 consecutive blocks of 1s and the block can be viewed in a circular fashion where the first row and the last row are adjacent rows. It has been proved that for a certain $k \geq 2$ the consecutive k-block problem remains NP-complete even if we restrict ourselves to symmetric matrices or matrices with at most 3 consecutive 1s blocks per row. This result shows that the cubic graph G admits a Hamiltonian circuit if and only if there exists a permutation of the rows for $B = A +1$ whose result is a matrix that has at most 2 consecutive blocks of 1s per column.

The complexity result shows that the Hamiltonian path problem remains NP-complete when restricted to triply connected, planar graphs with no less than 5 edges.

In what follows, we show the complexity results that we classify in the following two categories:

1- Consecutive linear blocks :

- ***2-consecutive blocks (symmetrical matrix)***

Instance: a symmetrical binary matrix M with at least 3 consecutive blocks of 1s on each row.

Question: Is there a permutation line for M that gives a matrix with at least 2 consecutive blocks of 1s per column?

- ***Consecutive blocks with matrix having 3 lines of 1s***.

Instance: a symmetrical binary matrix M with at least 3 consecutive blocks of 1s on each row and a constant . $k \geq 2$

Question: Is there a permutation line for M that gives a matrix with at least k consecutive blocks of 1s per column?

Consecutive blocks for symmetrical matrices.

Instance: a symmetrical binary matrix M and a constant . $k \geq 2$

Question: Is there a permutation line for M that gives a matrix with at least k consecutive blocks of 1s per column?

2- Consecutive circular blocks :

- ***2-Circular blocks (symmetrical matrix)***

Instance: a symmetrical binary matrix M with at least 3 circular blocks of 1s on each row.

Question: Is there a permutation line for M that gives a matrix with at least 2 circular blocks of 1s per column?

- ***Circular blocks with matrix having 3 lines of 1s.***

Instance: a symmetrical binary matrix M with at least 3 circular blocks of 1s on each row and a constant $k > 2$.

Question: Is there a permutation line for M that gives a matrix with at least k circular blocks of 1s per column?

- ***Circular blocks for symmetrical matrices.***

Instance: a symmetrical binary matrix M and a constant . $k \geq 2$

Question: Is there a permutation line for M that gives a matrix with at least k circular blocks of 1s per column?

We will now establish the relationship between the Hamiltonian circuit and the consecutive blocks on one side and the circular blocks on the other.

Consider the regular graph $G(V,E)$ with $g(G) \geq 5$. . There exists a Hamiltonian circuit if and only if the rows of the matrix $A(G) +1$ can be permuted to have at least k-1 circular blocks on each column. The proof is given in [17] which we summarise below:

Let $V = (v_1, v_2, \ldots\ldots\ldots\ldots v_n)$ and $B = A(G) + I$

Suppose the graph G admits a Hamiltonian circuit $v_{(i)\ (1)}$, $v_{(i)\ (2)}$, , v_{in}, $v_{(i)\ (1)}$and consider the permutations that rearrange the rows of B in the order $i_{(1)}$,i_2, i_n. With these permutations and for any column $_{ij}$, the adjacency of row ij_(1) (or $_{in}$ if ij - 1) and row $_{ij}$ as well as the adjacency of row $_{ij}$ and row $_{ij(+1)}$ (or $i_1$if ij - n), allows us to have the three consecutive 1s on column $_{ij}$ which are $B(ij\_{(1)}\,,_{ij})$, $B(_{ij,ij})$, $B(_{ij(+1)}\,,_{ij})$ B, so column $_{ij}$ has at least k - 1blocks (because each column has exactly $(k+1)1\ s$.)

A question can be raised in this context: what happens if we only use two lines next to each other? If $v_{(i)\ (1)}$is adjacent to $v_{(i)\ (2)}$in G , then putting line $i_1$adjacent to line $i_2$in B gives us two consecutive 1s : $B(i_{(1)}, i_1)$ and $B(i_{(2)}, i_1)$ on column $i_1$and two consecutive 1s $B(i_{(1)}, i_2)$ and $B(i_{(2)}, i_2)$ on column i_2.

But for any line $i3$ other than $i_1$and i_2, we also have $B(i_{(1)}, i_3) = 0$ where $B(i_{(1)}, i_3) = 0$. (Otherwise the circuit $v_{(1)(1)}, v_{(i)2}, v_{(i)3}, v_{(1)(1)}$is of length 3 which contradicts $g(G) \geq 5$).

So if now $v_{(i)(1)}$is adjacent to $v_{(i)(2)}$then rows $i_1$and $i_{(2)}$,which are adjacent in B , give us two pairs of adjacent 1s in all columns (one pair in column $i_1$and another pair in column $i_{(2)}$).

For the remainder of this section, we will begin by presenting the consecutive 1s problem for particular matrices and problems of interest to our study. The reader can consult the bibliography in the appendix for further research into the variants in this field. We will see that the problem of this type can be approximated for matrices (2,3) in which there are not two identical columns and can also be approximated for general matrices. We are also interested in the case of symmetric matrices.

To understand the problem properly, we will first take a particular variant of the problem of consecutive 1s. The case of a matrix in which an integer k is given. This is called COS (Consecutive ones submatrix).

The problem is formulated as follows:

Consider the problem of a binary matrix (0,1) whose instance is a matrix $A(m \times n)$ and a positive integer k. Is there a sub-matrix $B(m \times k)$ of A such that this matrix has the property of consecutive 1s, i.e. the columns of B can be permuted such that in each row all the 1s are consecutive.

However, the problem of consecutive 1s concerning a sub-matrix is NP-complete. This result is due to the fact that one of the variants of this problem is used to show that the other problems are also NP-Complete [18].

The proof of NP-completeness for this type of problem is given in [4]. But which proves the NP-completeness of a problem slightly different from the original problem (COS) as defined above.

This problem is posed by Booth [1] and was declared as a transformation of the Hamiltonian path problem , we will discuss this problem later. Theorem 1 from [4] shows.

Theorem 1: Given a matrix $M(m \times n)$, deciding whether there exists a sub-matrix $(m \times k)$ with the 1s consecutive property is NP-complete.

Proof: A simple reduction can be made from the Hamiltonian path. Let be the graph $G(V,E)$ and M the incidence matrix. G admits a Hamiltonian path if there are $n \times (n-1)$ sub-matrices

M' with the property of consecutive 1s. The Hamiltonian path problem is an NP-complete problem.

Booth's notes on this are clear, the incidence matrix M for G is nx m (n : number of vertices and m : number of edges), or M has a row for each vertex and a column for each edge with a 1 on the j' ^{ith}element if and only if vertex i is incident on edge j (noted that each column has exactly two 1s). If there is an $(n \times (n-1))$ sub-matrix M' with the property of consecutive 1s then there is a permutation P such that PM' verifies the property of consecutive 1s.

To see this, note that vertex i can only have an edge with vertex $(i-1)$ and $(i+1)$. If the columns have the property of consecutive 1s, then there are $(n-1)$ columns. This is the case where vertex 1 is connected to vertex 2, vertex 2 to vertex 3, etc.

However, the theorem above and the proof are not the proof of the consecutive 1s problem for sub-matrices (COS) given in [4]. In fact they are the proof of the NP-completeness of the problem of the form we will present below.

Let be the binary matrix $A(m, n)$ and a positive integer k .

Is there a sub-matrix $B(m, k)$ of A which has the property of consecutive 1s, i.e. the rows of B can be permuted such that in each column all the 1s are consecutive.

Note that in the COS problem, we swap the columns after selecting them and we would like to have the consecutive 1s property for the rows, but in the second problem, we swap the rows after selecting the columns and we would like to have consecutive ones for each column. These two problems are not equivalent. If$k \neq n$ (if $k = n$, then one of the problems is the transpose of the other and the two are symmetrical). The cause of the error in reality is that the original consecutive 1s property was put for columns (i.e., permutation of rows) and Theorem 1 presented in [4] only talks about the consecutive 1s property (i.e. it says nothing about rows and columns). This brings us to Theorem 2 in [4].

Theorem 2: The COS problem is NP-complete.

Proof: We reduce the undirected Hamiltonian path with degrees equal to 3 to a maximum for this problem and formulate it as an undirected graph G with the degree of each vertex less than or equal to 3. Now decide whether this graph G admits a Hamiltonian path or not as follows:

Let $G(V,E)$ be a graph with n vertices and m edges. Consider the graph G', an extension of G by adding an incident edge to each vertex v of G (the new adjacent vertex corresponding to v,is v').

The new graph $G'(V',E')$ now has $n' = 2n$ vertices and $m' = m + n$ edges. Consider the incident matrix M for G' (note that the incident matrix M for G' is $n' \times m'$ when M has a row for each vertex and a column for each edge with a 1 in $M(i, j)$if and only if vertex i *is* incident on vertex j) each column has exactly two 1s and each row has at least four 1s.

Assume that G has a Hamiltonian path $v_1, v_2, v_n,$ so the columns corresponding to edges (v_1,v'),$(v_1,v2)$,$(v2,v2)$,$(v2,v_3)$,$(v3,v')$, ,$(v_{(n)},v'_n)$ form a sub-matrix $(n'x\ (n1))$ with the 1s consecutive property for the rows.

If now G admits a Hamiltonian path and M admits a submatrix M' of dimension $(n' \times (n'-1))$ with the property of consecutive ones for the lines.

No selection of the k' columns of M can admit a cycle in G', otherwise these k' columns will have at least one row without the property of consecutive 1s. So this selection corresponds to a forest in G' and since it has n'-1 columns (edges), *30* it must be a tree T. Now let v be a vertex of the original set of vertices V, it is not possible to have more than 2 neighbours in T which also belong to the set of vertices V.

Let $v_{(1)}$,v_2, and $v_3$be the first three vertices (corresponding to the edges in consecutive order) and which are adjacent to v. $v2$, is a hanging edge but does not appear on either side, because the edge v,$v_1$is on the side of $v_2$and on the other side the edge v,v_3. So the tree corresponding to the Hamiltonian path in the graph G.

By this construction each column has at most two 1s and each row has at most four 1s, which leads us to the following corollary:

Corollary 1: The COS problem is NP-complete for matrices with at most two ones in each column and at most four ones in each row.

Let us now show another result using the theorem quoted in [4].

Theorem 3: For matrices that have at most two 1s in each column and at most two 1s in each row, the COS problem can be solved in polynomial time.

Proof: The graph G is constructed from the matrix A as follows.

A vertex is associated with each line and an edge is created between 2 vertices v and u if and only if the two lines corresponding to v and u have a "1" in the same column. The degree of each vertex in G is at most two, so its components are all cycles or paths.

Obviously, we can't select all the columns corresponding to the vertices of the cycle.

Let c' be the number of cycles in G and c the number of columns in the matrix. It is easy to see that the COS problem has a solution if and only if $k \leq c - c'$ Note that all the operations performed previously can be performed in polynomial time.

The complexity of this case, in which we have at most two ones in each column and at most three ones in each row, remains open. Note that this problem, where we ask whether B has the circular ones property for the rows, (the columns of B can be permuted so that on each row there will be an occurrence of "1" or "0" consecutively) is also NP-complete.

Therefore, it has been proven that the COS problem remains NP-hard for :

1- matrices (2,3) with a maximum of two "1" in each column and a maximum of three "1" in each row.

2- (3,2) matrices with a maximum of three "1 "s in each column and a maximum of two "1 "s in each row.

On the other hand, the COS problem can be solved in polynomial time for binary matrices with at most two "1 "s in each row and column.

Studying this problem (COS) for matrices with a small number of "1" in the columns or rows is not only interesting from a theoretical point of view but also from a practical one. Currently, restrictions on hollow matrices have been proposed by Greenberg and Istrail [19], which makes the physical layout problem simple and malleable, for example.

We will discuss here the case where the problem remains an NP-complete problem for the two cases mentioned above. However, it has been proved that the COS problem is 0.8 approximable for (2,3) matrices and 0.5 approximable for (3,2) matrices.

We will now give an overview of the algorithms presented in the article by Tan and Zhang [22].

1-Approximation algorithm (0,8) for matrices (2,3) :

Theorem 4: For a COS problem and for any matrix $A(2,3)$, there exists a polynomial approximation algorithm (0,8).

Proof: Given the matrix $A(2,3)$ with m rows and n columns, then $G(A)$ = (V,E) with m n nodes and n edges, so each row contains at most three "1 "s in A . Each node is of degree 3 at most.

For any subset $C \subseteq \{1,2,\ldots,n\}$, if $A(C)$ has the property of consecutive 1s, $G_{(c)}(A)$ is the union of caterpillar subtrees. However $G_{(c)}(A)$ has at most $(m-1)$ edges and therefore C contains at most $(m-1)$ columns.

To find the (0.8)-approximation solution to the COS problem of A , we need to find a spanning tree T in $G(A)$. A well-detailed algorithm is presented in [17], and it is easy to see that the algorithm takes $O(m)$ time. Note that the tree with at most 6 nodes is Caterpillar if each node is of degree 3 at most.

2-Approximation algorithm (0,5) for matrices (3,2) :

This algorithm was presented using Lovàsy's partition theorem [18] in graph theory to obtain a (0,5) approximation algorithm for the COS problem for (3,2) matrices. It is presented as follows:

A (3,2) matrix contains at least three ones in each column and at most two ones in each row. Other permutations of columns preserve the consecuity of the "1 "s in the row with at most one "1". We are focusing here on matrices (3,2) which leave exactly two "1s" on each row.

Let $A(3,2)$ be a matrix, so this defines only a graph $G(A)$ with degree 3 at most in which the nodes correspond one by one to the columns in A and the edges to the rows. Without loss of generality, let A have m rows and n columns and the

corresponding graph $G(A)$with a set of n nodes $V = \{1,2, \quad ,n\}$, plus

From this we have the following facts.

Lemma 1. Let $C = \{i1,_{i2}, \qquad _{ik}\}$ a subset of columns of A, the sub-matrix

$A(C)$ of A composed of the columns in C has the property of consecutive ones if and only if the subgraph $G(A)_C$induced by the subset of n nodes C ç V is the union of paths.

Proof: Consider that $A(C)$ with the consecutive ones property, be node i' e C, if there are 3 adjacent nodes ij $,_{ik},i_{l}$, then any permutation of C cannot leave consecutive "1 "s on the lines corresponding to $(i\ ',\ _{ij}Xi\ ',\ ik\ _{)}\ and\ (i\ ',\ )$ since there are only two "1 "s on each line, so each node in $G(A)_C$is at most of degree 2. By similarity, we can show that $G(A)_C$does not contain cycles, so the node that induces the subgraph is the union of paths and isolated nodes.

If now $G(A)\ _C$is the union of the paths, then if we order the set of all the columns on each path in the same order as they appear on the path, the resulting matrix has consecutive "1 "s on each row.

Using Lemma 1, a (0,5) approximation algorithm for the (3,2) matrix is presented.

In the following, we present another class of this type of problem which goes back to Kou [13]. We recall for this case that the consecutive 1s problem for binary matrices is decided if there exists an order of rows such that each column contains at most one block of 1.

This problem can be solved in polynomial time, and several problems of this type are interesting, such as the one presented by Goldberg and Al [20], which shows that the problem of k consecutive blocks of 1s is NP-Complete for a given $k \geq 2$.

This problem consists of reordering the rows of the matrix such that each column contains at most k blocks of 1s. It is shown that this problem is always NP-complete even for "Hollow" matrices with at most 8 ones in any row or column.

Kou has also shown that there is an NP-complete problem for determining the minimum number of partitioned columns (rows) required for a binary matrix in order to obtain the property of consecutive 1s. This consists of partitioning a column c and replacing it with two columns c' and c'' such that c has a 1 on row r if and only if c' or c'' has a 1 on the same row. This procedure is analogous and can be applied to cliques. This problem can be formulated in the same way as the 2-one consecutive problem.

Given a matrix M, the question is whether each column of M can be partitioned into at most two columns so that the resulting matrix has the 1s consecutive property.

Given a binary matrix, the question is when can each row of M be partitioned into at least two rows such that the resulting matrix has the property of consecutive 1s. This problem is NP-complete even for hollow matrices with at most 6 ones in any row or column.

The problem of consecutive 1s with a Partitioned row is formulated in the following way . Let be a binary matrix M . Is there a row r of M that can be replaced by at most 2 rows r' and r'' such that for each column c, row r has a 1 in column c if and only if r' or r'' has a 1 in column c and such that the result matrix has the property of consecutive 1s?

Corollary 2. The problem of consecutive 1s with a partitioned row is also NP-complete for hollow matrices with at most six 1s per row and column.

The k 1s consecutive problem is defined by a binary matrix M . We answer the question: Is there an order of the rows of M such that each column contains at most k blocks of 1?

The case $k = 2$ is of particular interest, as we know that the k-$1s$ consecutive problem is NP-Complete (Goldberg et Al [20]) for each $k \geq 2$. They ask the question when does this problem remain NP-Complete when the matrix is of the hollow type? They only showed that the problem remains NP-Complete if each column contains at most a constant number of 1's for $k \geq 3$.

We see here that the problem remains NP-complete for hollow matrices and $k = 2$. To do this we will show that the version of the k -1s consecutive problem is NP-Complete for hollow matrices and $k = 2$, we will see that the rows of the matrix are such that they form a circular alignment and so if a 1 occurs in the first and last rows, automatically they are in the same block.

A second class, which goes back to Kou [13], can also be posed in the form of an augmented matrix. We characterise a large class of simple algorithms that do not find near-optimal augmentation of their input matrices. Kou has shown that there are matrices for which these algorithms find augmentations that are even far from optimal. This type of problem is raised for applications in network improvement, which consists of adding new links to a given network in order to improve its performance.

This problem can be modelled as a graph augmentation problem as follows: given a graph G and an integer k, we want to add a minimum number of edges to the graph G so as to obtain an augmented graph of diameter less than or equal to k .

Note that the complementary problem of the Hamiltonian path is to find a set of augmented edges such that the augmented graph admits a Hamiltonian path. We show that this path is far from having an approximation algorithm *36* with a constant ratio. This problem remains difficult to approximate even if the subgraph is a tree. However, if the weights of the edges are restricted (between 1 and 2), the complementary problem of the Hamiltonian path is still NP-hard. When the tree is now a k-tree, we can find an approximation algorithm with a performance ratio equal to 1.5.

Next, and in the same vein, Kou studies another variant of this problem which consists of minimising the record space (minimising redundant records). He then studies the problem of minimising consecutive blocks of records. These two versions of the problem can take the following form:

Problem 1: Minimising redundant records :

Let be the matrix $A(m,n)$ and let be $Q_j = \{r_i | b_{ij} = 1\}$ for $1 \leq j \leq m$

The question is to find the length of the sequence x in the alphabet $R = \{r_1, r_2,,,,,,,, r_n\}$ such that the elements of Q_j appear consecutively in x, for $j = 1,2,,,,,,,m$.

Problem 2: Minimising consecutive blocks of appropriate record storage.

Given the matrix $A(m,n)$, find a permutation of A such that the total number of consecutive blocks of 1's in the column of A is minimised.

These two problems are NP-Complete, which means that if an efficient algorithm can be found to solve one of these two problems, several other problems (e.g. Hamiltonian path problem, Travelling Salesman) may also have efficient algorithms and solutions.

The Hamiltonian path is defined here as being a sequence of n distinct vertices, the cost of the Hamiltonian path in a graph is the sum of the costs of the edges of the path, the following lemma presented in the artice of L.T. Kou (13] shows the relation between the cost of the Hamiltonian path and the total number of consecutive 1s.

Lemma 2. Given the matrix $A(m,n)$, the cost of the Hamiltonian path $(1,2, ,n)$ is k if and only if the total number of consecutive blocks of 1s on the column of A is $k + c$ or c is the number of 1s on the n'i^{th}row of A .

Proof: Let $N_,$be the total number of consecutive blocks of1s in the columns of A and $N_i$the total number of consecutive blocks of 1s at the end of row i of A . Clearly :

$$N = N_1 + N_2 + ... + N_n$$

and it is also clear that for $1 < i < n$, $N = K$ in A if and only if the cost of

$$f([i, i+1]) = K_i$$

On the other hand, $N_n = c$.

So the cost of the Hamiltonian path $(1,2, n)$

$$= f([1,2]) + f([2,3]) + ... + f([n-1, n])$$
$$= N_1 + N_2 + ...N_{n-1} = N - c$$

and the proof is complete.

The version presented by Kou is formulated around this problem, the author denotes the total number of blocks of consecutive 1s on the columns by the question. Is there a permutation matrix P such that $PA = k + c$, where c is the number of 1s in the n' th row of PA .

We end this section by giving the definition of the circular k -1-consecutive problem:

Instance: A binary matrix M

Question: Is there an order of the rows of M such that each column contains at most k blocks of 1s, if it is assumed that the columns are circular and the first and last elements are neighbours?

Finally, here are the complexity results we have established for the CBM problem. We have identified the following points.

By studying the complexity of the problem and the approximation paths on particular matrices, we have established a new approximation proof for the problem of minimising the number of consecutive blocks of 1s. Two results will be presented. One result states that CBM is NP-complete even when restricted to binary matrices with two 1s per row. A new result is presented as a proof that CBM is 1.5 approximable.

Chapter 2

A new negative result: CBM is NP-complete even when the binary matrix has only two 1s per row

2.1 Introduction.

Our goal in this chapter is to prove that CBM is NP-complete even on special instances such as binary matrices with only two 1's per row. On the other hand, we consider the problem of the Hamiltonian chame of maximum weight, but restricted to instances with a weight equal to 1 or 2. We then propose a polynomial transformation which reduces any instance of the second problem to an instance of the first, i.e. CBM. This shows that CBM is at least as difficult as the first. Since the first problem is NP-complete, it follows that CBM is also NP-complete.

Given a binary matrix A, let $A(\pi)$ be the matrix induced by a permutation π of the columns of A. The term $n(A(\pi))$ denotes the number of coefficients $a_{in(j)})$ such that $\alpha_{i\pi(j)}) = 1$ and ($a_{i..(0\,j(+1))}$= 0 or $j = n$) . The CBMP problem consists of looking for a permutation π^* such that $n(A^*,) = \min_{(\pi)} n(A(\pi))$.

2.2 2CBM problem.

Consider the restriction of the CBM problem, called 2CBM, to matrices with two 1s per row. Consider also the restriction of the maximum weight Hamiltonian chain problem called 2MWH, where the weights of the edges have a weight of 1 or 2. The 2MWH problem is NP-hard as long as the Hamiltonian chain problem can be directly polynomially reduced to 2MWH.

2.3 Maximum Hamiltonian path weight problem .

Seeking to minimise or maximise the weight of data structures such as trees, paths, cycles or sets of vertices is important in studies of Hamiltonian path finding. The weight of such a structure is then naturally defined by the sum of the weights of its components (in general, vertices or edges). However, it is sometimes more interesting to define the weight as being equal to the weight of the smallest (or largest) component.

To give an example, if we model a network as a graph, where the weights of the different edges correspond to the maximum flow they can let through, then we would like the weight of a path to also correspond to the maximum flow it can let through. Still in the context of links in a network, we can look for the fastest path, i.e. the path with the greatest weight, between two nodes.

The weight of a path from vertex 1 to vertex n via vertices 2, 3, ..., $n-1$ should be calculated simply as a function of the weight of path 1, 2, 3, ..., n - 1 and the weight of edge (n -1,n). Naturally, the weight of a path must increase when an additional edge is added.

Instance: The graph G.

Question: Is there a chain that passes through all the vertices exactly once?

Such a path is called a Hamiltonian chain, or a Hamiltonian path. If the path is now closed, it is called a Hamiltonian cycle. A Hamiltonian cycle is an elementary cycle containing all the vertices of the graph.

This graph will be oriented and valuated, with each of its edges having a direction and a valuation. In this case, the problem involves finding a Hamiltonian path of maximum weight.

Many algorithms have been developed to solve problems involving graphs. For this reason, it has become natural to transform any problem into a graph problem in order to use them.

In this case, the consecutive blocks problem results in the search for a Hamiltonian path of maximum weight in this graph. Consequently, this arrangement is translated into the graph by the search for a path passing through all the vertices and whose weight is maximal. So the version of the Hamiltonian path problem in an undirected graph corresponds exactly to the travelling salesman problem that we will see in Chapter 3.

2.4 From the 2MWH problem to the 2CBM problem.

The transition from the 2MWH problem to the 2CBM problem consists in making a restriction on the original problem by formulating the problem based on a restriction of the weights of the edges which are of degrees 1 or 2, we have noted that this problem remains NP-hard. We will in fact study the version of the problem where the weights on the edges connecting the cities verify the triangular inequality of Christofides [25]. Without this restriction, the problem cannot be approximated.

First of all, it is worth highlighting the links between the problems we are interested in and the Travelling Salesman Problem (TSP). Firstly, there are a large number of publicly available algorithms for solving large instances of the TSP. Moreover, it is capable of providing the optimal solution on instances of up to several hundred nodes. Another advantage of such a reduction comes from the theoretical knowledge accumulated on the TSP. For example, it is known that on graphs respecting the triangular inequality, the TSP is 1.5 approximable, which means that there is an algorithm (described by Christofides) and in $O(n^3)$ capable of generating a solution less than 1.5 times the optimal solution. This result is of course valid for the TSP graph.

We need to recall the definition and some properties of the travelling salesman problem (TSP), and we will show how this problem can be useful for understanding and solving our problem. In particular, we will show that when there is a restriction on weights, the problem is equivalent to a certain variant of the TSP. The travelling salesman problem thus defined assumes that all the vertices of the graph are connected (complete graph).

This is represented by a matrix with two 1s per row following the polynomial reduction performed on the original problem.

2.5 Polynomial transformation :

We would like to show that the 2MWH problem reduces polynomially to the 2CBM problem. To do this, we need to :

(i) Present a transformation that shows that each instance of the 2MWH problem is an instance of the 2CBM problem;

(ii) Show that it is polynomial,

(iii) Prove that the Hamiltonian path π (in the graph that it is an instance of 2MWH) has maximum weight if and only if it induces an order π of the columns da the binary matrix (as an instance of the 2CBM problem obtained via this transformation) that minimises the number $n\left(A(\pi)\right)$.

(i) Let's start by giving the transformation:

Procedure *transformation.*

Input : *nxn-matixW* such that $w_{ij} = 0$, et $w_{ij} = w_{ji} \in \{1,2\}, i \neq j$

Output : *r x n - binary matrix A* with two1 per line /* *r* to be defined*/

$r \leftarrow \sum_{i=1}^{n-1}\sum_{j=i+1}^{n} w_{ij}; t \leftarrow 0;$

$for\, i=1 \text{ to } r\, do \quad for\, j=1 \text{ to } n\, do \;\; A_{ij} \leftarrow 0;$

$for\, i=1 \text{ to } n-1\, do \quad for\, j=i+1 \text{ to } n\, do \;\; for\, k=1\, to\, w_{ij}\, do;$

$begin\; t \leftarrow t+1;$

$A_{ti} \leftarrow 1;$

$A_{tj} \leftarrow 1$

end

end.

By this construction, A is binary and has precisely two non-zero elements per line.

Example 1. The procedure consists of transforming the matrix W into another matrix A .

We have $r = 4$. This value is obtained by the formula described above.

$$W = \begin{pmatrix} 0 & 1 & 2 \\ 1 & 0 & 1 \\ 2 & 1 & 0 \end{pmatrix}$$

Transformation

$$A = \begin{pmatrix} 1 & 1 & 0 \\ 1 & 0 & 1 \\ 1 & 0 & 1 \\ 0 & 1 & 1 \end{pmatrix}$$

(ii) The complexity of the transformation dominated by the dimension of A is

of order $O(n^{3)}$, so clearly : $n(n-1)/2 \le r \le n(n-1)$.

Proposition 1. The matrix $W = A^{(T)\,A}$, except for the diagonal elements.

Let's calculate :

$$W = A^T A$$

$$= \begin{pmatrix} 1 & 1 & 1 & 0 \\ 1 & 0 & 0 & 1 \\ 0 & 1 & 1 & 1 \end{pmatrix} * \begin{pmatrix} 1 & 1 & 0 \\ 1 & 0 & 1 \\ 1 & 0 & 1 \\ 0 & 1 & 1 \end{pmatrix} = \begin{pmatrix} 3 & 1 & 2 \\ 1 & 2 & 1 \\ 2 & 1 & 3 \end{pmatrix}$$

Proof: Given w_{ij}, from the assignment made in the 3^{th}nested loop, it follows that the scalar product $a_i^T \times a_j$. where a_i and a_j are two distinct columns of the matrix A, must be $\ge w_{ij}$.

Now let us observe that by this construction, the rows of the matrix A are distinct, except for the pair which results from the case where $w_j = 2$.

From the last observation, and since each row of the matrix A contains two non-zero coefficients, the scalar product $a_i^T \times a_j$ must be equal to w_j.

There is a one-to-one correspondence between the Hamiltonian paths in the complete graph on n vertices **with the weight of the edges of the matrix** W and the order of the columns of A .

Proposition 2. Let the following form represent the number $2r$ as defined in the algorithm.

$$n(A_\pi) + \sum_{j-1}^{n-1} W_{\pi(j)\pi(j+1)} = 2r, \forall \pi.$$

Proof: The value $2r$ is the number of 1s in the matrix A that do not depend on π. In order to obtain the number $n(A_{(\pi)})$ of coefficients $a_{(in)\ (j)}$ such that $a_{(in)\ (j)} = 1$ and $(a_{i''(()\ j+1)}) = 0$ or $j = n)$, we need to subtract from $2r$ the number of coefficients $a_{(in()\ j)}$ such that $a_{(in)}\ (j\ ) = 1$ and $a_{(i\pi)\ (()\ j(+1))}$**= 1**, this is done by subtracting the term :

$$\sum_{j=1}^{n-1} W_{\pi(j)\pi(j+1)} = \sum_{j=1}^{n-1} \sum_{i=1}^{m} a_{i\pi(j)} \times a_{i\pi(j+1).}$$

So

$$n(A_\pi) = 2r - \sum_{j=1}^{n-1} W_{\pi(j)\pi(j+1)}$$

(iii) Proposition 2 shows that the order π of the columns of the matrix A minimises the number $n(A^*)$if and only if π is a Hamiltonian path of maximum weight.

2.6 Conclusion

An immediate consequence of **the polynomial reduction is that the 2CBM problem is as difficult as the 2MWH problem.**

Theorem 1: The 2CBM problem is NP-hard.

Example 2: in the previous example, we have $\pi = (1,2,3)$, $n(A_n) = 6$, and the weight of π is 2. We have $6+2=2r$ =8. Therefore, we verify that a minimal column order of the matrix A is π^* $= (1,3,2)$ giving $n(A_{\pi^*})$ =5. . The weight of the corresponding path (maximum Hamiltonian weight) is 3. We still have 5+3=8.

We have $r = 4$ as calculated in the transformation (order 1,2,3)

$$n(A_\pi) = 2r - \sum_{j=1}^{n-1} W_{\pi(j)\pi(j+1)}$$
$$= 8 - 2 = 6$$

$$A = \begin{pmatrix} 1 & 1 & 0 \\ 1 & 0 & 1 \\ 1 & 0 & 1 \\ 0 & 1 & 1 \end{pmatrix}$$

The minimum order is obtained by (1,3,2) with $n(A_{\pi^*}) = 2r - \sum_{j=1}^{n-1} W_{\pi(j)\pi(j+1)}$

and with a weight = 3 hence 5+3=8

$$A_{\pi^*} = \begin{pmatrix} 1 & 0 & 1 \\ 1 & 1 & 0 \\ 1 & 1 & 0 \\ 0 & 1 & 1 \end{pmatrix}$$

Chapter 3

Another positive result: MINIMISE THE NUMBER OF BLOCKS OF 1 CONSECUTIVE is 1.5 approximate.

3.1 Introduction

Let be a binary matrix A and $A(\pi)$matrix result of a permutation π of the columns of A . We will recall the formal definition of the following three combinatorial optimisation problems.

(i) Minimising the number of consecutive 1s blocks (CBM)

Instance: positive integers m,n,k and the binary matrix $A(mxn)$

Question: is there a permutation π of $1,...n$ such that the number of

coefficients A_j^i of $A(\pi)$, verifying $A_j^i = 1$ and $j = n$ or $A_{j+1}^i = 0$, be equal to k?

(ii) Minimising the number of blocks of consecutive circular 1s (CIR) :

Instance: positive integers m,n, K and $A(m \times (n+1))$ a binary matrix with columns $0,1, \quad , n$

Question: Is there a permutation π of $0, \; n$ such that the number of coefficients A_j^i of $A_\pi$verifying $A_j^i = 1$ and either $A_{j+1}^i = 0$ or ($j = n$ and $A_0^i = 0)$, is equal to K?

It is clear that this number is the same as the one that counts the coefficients A_j^i such as $A_j^i = 0$ and $A_{j-1}^i = 1$ or $(j = 0 \; et \; A_n^i = 1)$

(iii) Travelling Salesman Symmetrical (TS) :

Instance: positive integers n,L (there are n +1 cities denoted $0,1,...n$) and a symmetrical distance matrix $\{D_j^i\} 0 \le i, j \le n$

Question: Is there a route between towns such as :

$$\sum_{j=0}^{n-1} D_{\pi(j+1)}^{\pi(j)} + D_{\pi(n)}^{\pi(0)} = L \; ?$$

The aim of the CBM optimisation problem is to find a permutation of the columns of A that gives the minimum number of coefficients quoted in the definition of the CBM problem.

There is an interesting question that has not yet been resolved. Can the CBM optimisation problem be approached with a proof of guaranteed performance? [24].

To formalise this question, let A be a CBM instance and let $opt(A)$ be the minimum number of desired coefficients of A obtained from a certain optimal permutation of its columns. Is there a polynomial-time heuristic H and a constant $\alpha, 1 \prec \alpha \prec \infty$ such that for all instances A, $H(A) \leq \alpha.opt(A)$ where $H(A)$ is the number of each coefficient produced by H applied to A?

We will give a positive answer of "yes", with a certain $a = 1.5$. We proceed as follows. First we will polynomially transform the CBM problem into a CIR problem, then a second transformation from the CIR problem to the TS problem which obeys the triangular inequality.

Since these two transformations are approximately preserved and since we know that Christofides' minimal spanning tree heuristic [25,26] is 1.5 approximate for the TS problem. So it is **a heuristic for the CBM optimisation problem**. (Sequential application of the two transformations followed by the Christofides heuristic)

Procedure:

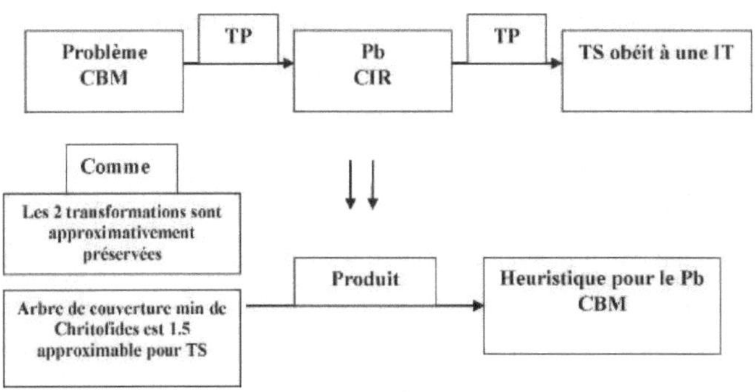

3.2 ε - Approximation

Approximation algorithms are used to find approximate solutions to optimisation problems, and are generally associated with NP-hard problems by the absence of efficient algorithms for solving these problems in polynomial time. On the other hand, reasonable solutions can be found in polynomial time, with provable quality, i.e. with a constant factor very close to the optimal solution. Approximation algorithms are widely used for problems for which polynomial-time algorithms are known but which become very expensive due to the size of the data.

An approximation algorithm is a heuristic that guarantees that the quality of the solution provided is less (if minimised) than a constant. In relation to the optimal quality of a solution, for all possible instances of the problem.

This is the case, for example, with the Commercial Traveller, where the weights satisfy triangular inequalities because the minimum weight of a spanning tree is always less than twice the optimum.

3.3 The travelling salesman (TS) problem :

A commercial traveller has to visit a finite number of towns once and then return to his point of origin. The principle is to find the order in which the traveller visits the cities that minimises the total distance travelled. This combinatorial optimisation problem belongs to the class of NP-complete problems because of its complexity and the large number of problems derived from it.

As a restriction to this problem, find the traveller's route so that he visits all the cities within the maximum journey time.

This problem can be modelled using a graph made up of a set of vertices and a set of edges. Each vertex represents a city, an edge symbolises the passage from one city to another, and it is associated with a weight that can represent a distance, a journey time or a cost.

To solve this problem, we need to find a cycle in this graph that passes through all the vertices once. Such a cycle is called Hamiltonian and is of minimal length.

Since there is an edge between each pair of vertices, the graph is said to be "complete". For any graph, a weight matrix can be established.

The Christofides algorithm is often associated with the travelling salesman problem for a number of reasons. The first is the combination of the minimum weight tree with the solution of the minimum weight cover problem. The second is Christofides' heuristic, which consists of finding the solution to an instance of the travelling salesman problem that satisfies the triangle inequality.

Christofides' algorithm is one of the first approximation algorithms and is responsible for the design of several approximation algorithms as a practical approach to many difficult problems.

In a special case, the distances between the cities are all 1 or 2, (and so the triangular inequality is satisfied) so there is a polynomial approximation algorithm that finds a more or less optimal route.

3.4 Heuristics based on the minimum weight tree

Obviously, several algorithms create a spanning tree. We are going to show that it is indeed of minimal weight. If we consider an optimal solution, let **edge-max** be the edge of maximum weight. Let's show that when all the edges with a weight less than that of "**edge-max**" have been considered, the spanning tree has been found.

The edges of the optimal spanning tree link all the vertices into a connected component. Now, having considered an edge in the optimal spanning tree, we know that its two ends are connected: either the edge has been added to the tree under construction, or it would have created a cycle. In other words, the two ends in question were already part of the same related component.

So we have a spanning tree when we consider 1'**edge-max**. All the edges of the tree in question are of weight less than or equal to **edge-max**. The tree is therefore of weight less than or equal to the tree of minimum weight. It is therefore done a spanning tree of minimum weight.

The principle of the open tree with minimum weight is as follows:

Given an undirected, connected graph, a **spanning tree** of this graph is a subset that is a tree and connects all the vertices together.

A graph can have several different spanning trees. We can associate a weight with each edge, which is a number representing the cost of that edge, and take the sum of the weights of the edges in the spanning tree. A **spanning tree with a minimum weight** is one whose weight is less than or equal to that of all the other spanning trees in the graph.

The minimum spanning tree is also known by other names, such as *minimum spanning tree* or *minimum subtending tree*.

Given a connected, undirected graph, a spanning tree is a sub-graph which is a tree whose vertices are connected. A graph can have different spanning trees. We can also assign weights to each edge to calculate the sum of the weights of the edges in the spanning tree. The minimal spanning tree or minimum weight tree is therefore a spanning tree with a weight less than or equal to the weight of the other spanning trees.

Given an optimal cycle, we eliminate the maximum length to obtain a spanning tree that is always less than or equal to the optimal. This is done by constructing an approximation which is limited to constructing a tree of minimal length, duplicating the edges, and then obtaining a cycle of length less than twice the optimal length, but where the vertices appear several times.

Thanks to the triangular inequality, all we have to do is traverse the vertices, jumping to the next vertex when we encounter one that has already been visited. We then obtain a cycle whose length is less than twice the optimal length, where the vertices only appear once. So if the weights of the edges respect the triangular inequality, then we can construct a spanning tree of minimal weight by eliminating multiple visits to the vertices from the path. In principle, this procedure gives us an approximate method for the travelling salesman.

Among the heuristics used to obtain approximate solutions to the travelling salesman problem is the nearest neighbour method. This is the most naive method for constructing a Hamiltonian path. As we mentioned earlier, we start from a vertex and at each step we move on to the nearest vertex not yet visited. The Christofides heuristic is similar to the method of calculating a spanning tree of minimum weight and extracting the circuit obtained by traversing this tree. However, it guarantees a better bound. In fact, the weight of a route constructed by this heuristic is at most 1.5 times the weight of an optimal route.

This heuristic also starts by calculating a spanning tree of minimum weight. It then constructs a coupling of minimum weight on vertices whose degree in the spanning tree is odd.

Let T be the set of edges of the spanning tree and the coupling. We see that T contains a circuit which is not necessarily simple, but which passes through all the vertices. We then extract a simple path by connecting the vertices in the order in which we encounter them by following the edges of T and skipping the vertices already visited.

3.5 Approximation 1.5 of the CBM optimization problem

Definition 1: A block of consecutive 1s on a row i of a binary matrix A is a sequence of sequentially occurring 1s. Formally it is a sequence of coefficients $A_p^i, A_{p+1}^i, \ldots\ldots, A_q^i$ on row i satisfying the following:

i) $A_j^i = 1, p \leq j \leq q$

ii) ou bien $p = 0, \text{ou } A_{p-1}^i = 0$

iii) ou bien $q = n, \text{ou } A_{q+1}^i = 0$

Clearly there is a one-to-one correspondence between the coefficients mentioned in the previous definition of the CBM problem and the consecutive blocks. So the question of the CBM problem can be rephrased as follows:

Question: Is there a permutation π *of* $1,\ldots n$ such that the number of blocks consecutive in $A(\pi)$ is k?

Similarly, the circular block on line i *is* either a consecutive block or a concatenation of two consecutive blocks. One starts with $A_0^i = 1$ and the other ends with $A_n^i = 1$ (which we call 2-CB).

There is also a one-to-one correspondence between the coefficients quoted in the definition of the CIR problem and the circular blocks So the question of the CIR problem becomes: Is there a permutation π *of 0,...n* such that the number of blocks

in A_π is K?

Proposition 1. CBM $\propto$ CIR ($\propto$: Reduction-preserving polynomial-time approximation).

Proof:

Let the binary matrix A be an instance of the CBM problem.

Let's calculate the instance B of the CIR problem by simply concatenating to A a column labelled "0" with zero coefficients.

$$B = [0|A]. \tag{3.1}$$

Let us now show that there exists a permutation π of n columns of A such that the number of consecutive blocks in $A(\pi)$ is k if and only if there exists a permutation Π of the n +1 columns of B such that the number of circular blocks in B_Π is $K = k$

Suppose there is such a permutation π and $A(\pi)$ has k consecutive blocks. Consider Π = $(0,\pi)$ such that $B_\Pi = [0|A]$. with k circular blocks which are the k consecutive blocks of $A(\pi)$, since there cannot be 2 consecutive blocks in B_Π.

Conversely, let Π= $(\Pi(0),\Pi(1), \quad ,\Pi(n))$ a permutation such *that*B_n

has k - circular blocks, there exists an integer s such that $\Pi(s) = 0$.

Now, any circular change of Π, in particular $(0,\Pi(s+1), \quad ,\Pi(n),\Pi(0), \quad ,\Pi(s-1))$ gives the same number k of blocks

circular. So we have $\pi = (\Pi(s+1), \quad ,\Pi(n),\Pi(0), \quad ,\Pi(s-1))$ such that

so that $A(\pi)$ has k consecutive blocks.

Proposition 2. CIR $^\curlyvee$TS

Proof: We need to define a "distance" between two columns of the matrix B obtained from the transformation (3.1), which expresses a kind of attraction between them: the smaller the distance between two columns, the greater our desire to put them side by side.

Let be the symmetric square matrix of order n +1

$$D = B^T (U - B) + (U - B)^T B \tag{3.2}$$

Where U is a matrix of order m x $(n+1)$, whose coefficients are 1s. The matrix D can be seen as an instance of the TS problem representing the matrix of distances between n +1 cities, denoted by$(0,1,....n)$. Since there is a one-to-one correspondence between the circuits and the circular permutations of the columns $of B$, we can freely use either of the two terms. We show that there exists a permutation π of the $(n+1)$ columns $of B$ whose result is a matrix $B(\pi)$having K circular blocks if and only if there exists a circuit of lengthL $= 2K$.

Suppose there exists a permutation π such that $C = B(\pi)$with K circular blocks.

Or : $E = C^T (U - C) + (U - C)^T C$

We would like to show that

$$L = \sum_{j=0}^{n-1} E^j_{j+1} + E^0_n = 2K$$

Let $C_j$be the jthcolumn of C . We have :

$$L = \sum_{j=0}^{n-1} [C_j^T (1 - C_{j+1}) + (1 - C_j)^T C_{j+1}]$$
$$+ C_0^T (1 - C_n) + (1 - C_0)^T C_n$$

$$= \sum_{j=0}^{n-1} \sum_{i=1}^{m} [c^i_j (1 - c^i_{j+1}) + (1 - c^i_j) c^i_{j+1}]$$
$$+ \sum_{i=1}^{m} [c^i_0 (1 - c^i_n) + (1 - c^i_0) c^i_n].$$

Where **1** is an m - vector whose coordinates are all 1's. Any multiplicative term in the last sum is binary, so those that verify one of the possibilities will be positive.

$C^i_j = 1$ et $C^i_{j+1} = 0$, ou

$C^i_j = 0$ et $C^i_{j+1} = 1$ pour $j, 0 \leq j \leq n-1$, ou

$C^i_0 = 1$ et $C^i_n = 0$ ou $C^i_0 = 0$ et $C^i_n = 1$

From the definition of the CIR problem and the remark that followed, the value of L is exactly twice the number of circular blocks in C .

Conversely, suppose there is a circuit π over the $(n+1)$ cities of length L with

$$L=\sum_{i=1}^{m}\left(\sum_{j=0}^{n-1}(C_j^i(1-C_{j+1}^i)+(1-C_j^i)C_{j+1}^i)+(C_0^i(1-C_n^i)+(1-C_0^i)C_n^i)\right)$$

Let us first show that the length L is even. In fact, for a fixed i the sum :

$$L_i=\sum_{j=0}^{n-1}\left[(C_j^i(1-C_{j+1}^i)+(1-C_j^i)C_{j+1}^i+(C_0^i(1-C_n^i)+(1-C_0^i)C_n^i)\right] \quad (3.3)$$

is even.

Note that for 3 binary numbers *a,b,c* , the sum.

$$a(1-b)+(1-a)b+b(1-c)+(1-b)c+a(1-c)+(1-a)c \quad (3.4)$$

As is.

$$a(1-b)+(1-a)b+b(1-c)+(1-b)c-a(1-c)-(1-a)c \quad (3.5)$$

Is even (whether 0 or 2 is simple to check).

The proof that the sum (3.3) is even is given by induction onn .

From (3.4) this is also true for $n=2$.

Let's show that this is true for n provided that it is true for n -1.

$$\sum_{j=0}^{n}(C_j^i(1-C_{j+1}^i)+(1-C_j^i)C_{j+1}^i)+(C_0^i(1-C_{n+1}^i)+(1-C_0^i)C_{n+1}^i)=$$

$$\left[\sum_{j=0}^{n-1}(C_j^i(1-C_{j+1}^i)+(1-C_j^i)C_{j+1}^i)+(C_0^i(1-C_n^i)+(1-C_0^i)C_n^i)\right]$$

$$+\left[C_n^i(1-C_{n+1}^i)+(1-C_n^i)C_{n+1}^i)+C_0^i(1-C_{n+1}^i)+(1-C_0^i)C_{n+1}^i)-C_0^i(1-C_n^i)-(1-C_0^i)C_n^i)\right]$$

The term between the first parenthesis is the same by induction hypothesis. The term between the second parenthesis is the same as in (3.5). So $L_i=2K_i$ for some integer K_i.

It remains to show that row i of $C=B_n$contains $K_i$circular blocks.

From (3.1) line i must contain a zero coefficient. If the C_j^i are 0, then $L_i=K_i=0$, so there are no circular blocks on line i.

Assuming that some C_j^is are 0s and others are 1s, let's see if the coefficients $C_j^i, 0\leq j\leq n$ can be placed sequentially on the circle so that C_0^i and C_n^i are adjacent.

Now the following procedure identifies all the circular blocks on line i .

Procedure for identifying circular blocks on line i .

1- Find the 1[st]occurrence 01 on the circle, (0 followed by 1 on the circle). Continue in a clockwise direction to find the 1[st]occurrence 10, (these two occurrences must exist and the 1 in both occurrences can be the same). These two pairs 01 and 10 are coupled and define the start and end of the circular block. They contribute the value 2 to the sum L_i.

2- remove all the 1's between the two 0's from the occurrences found, then put $L_i = L_i - 2$, or ($K_i = K_i - 1$).

3- repeat steps 1 and 2 until $L_i = 0$.

Proposition 3. The traveling salesman (TS) problem with a distance matrix D defined in (3.2) obeys the triangle inequality.

Proof: Given j_1, j_2, j_3, arbitrary cities. We must have :

$$D_{j2}^{j1} \leq D_{j3}^{j1} + D_{j3}^{j2} \qquad (3.6)$$

We have :

$$D_{j_2}^{j_1} = B_{j_1}^T(1 - B_{j_2}) + (1 - B_{j_1})^T B_{j_2}$$

$$= \sum_{i=1}^{m} [B_{j_1}^i(1 - B_{j_2}^i) + (1 - B_{j_1}^i)B_{j_2}^i]$$

$$= \sum_{i=1}^{m} \left[B_{j1}^i(1 - B_{j2}^i) + (1 - B_{j1}^i)B_{j2}^i\right]$$

And

$$D_{j3}^{j1} + D_{j3}^{j2} = \sum_{i=1}^{m} \left[(B_{j1}^i(1 - B_{j3}^i) + (1 - B_{j1}^i)B_{j3}^i) + (B_{j1}^i(1 - B_{j3}^i) + (1 - B_{j1}^i)B_{j3}^i)\right]$$

So inequality (3.6) is true since for any fixed i, we have :

$$B_{j1}^i(1 - B_{j2}^i) + (1 - B_{j1}^i)B_{j2}^i \leq B_{j1}^i(1 - B_{j3}^i) + (1 - B_{j1}^i)B_{j3}^i) + B_{j1}^i(1 - B_{j3}^i) + (1 - B_{j1}^i)B_{j3}^i$$

Because for any binary number: a,b,c, we have :

$$a(1-b) + (1-a)b \leq (1-c) + (1-a)c + b(1-c) + (1-b)c$$

Given a binary matrix A , consider the following heuristic for the CBM optimisation problem.

(i) Apply the transformation (3.1) to obtain the matrix B .

(ii) Apply (3.2) to obtain an instance D of TS (which obeys the inequality triangular of proposition 3).

(iii) Finally, apply the Christofides heuristic to instance D.

Clearly, this heuristic for the CBM optimisation problem is polynomial, which we summarise as follows:

- Let L be the length of the approximation circuit. According to Proposition 2, this circuit is a permutation of $n+1$ columns of B giving $k = L/2$ circular blocks (L is even).
- According to Proposition 1, if we remove city 0 from the circuit, we obtain a permutation (which is a chame over the n remaining cities) corresponding to a matrix with k consecutive blocks.
- Since the two transformations above preserve the approximation ratios, we can formulate the following result.
-

Theorem: The heuristic constructed for the CBM problem is 1.5 approximable [34].

Conclusion

After presenting some complexity problems for classes well known in combinatorial optimisation, we defined three problems, which are a restriction of the consecutive 1s problem. The consecutive blocks of 1s problem, the circular blocks problem and the travelling salesman problem.

We have proved with a number of polynomial transformations that the CBM problem is an NP-complete problem with its two versions. (CBM and 2CBM)

We have also established two important results in this thesis, on which we have established a polynomial-time heuristic for solving the consecutive block minimisation problem, which generates a solution to be 1.5 approximate. We have shown that this problem is NP-complete even when restricted to binary matrices with two 1s per row. However, it is possible to find a heuristic with better performance.

Remember that the symmetric matrix D, the instance of the travelling salesman's problem, is of course not ordinary, it obeys the law of triangular inequality, but it also has two new properties.

1- Each coefficient of the matrix D must not exceed the number m of rows of the matrix A .

2- Each cycle (not necessarily Hamiltonian) has an even length.

The question remains open in this context: do the two properties allow us to do better?

Bibliography

1- Kellogg Speed Booth (1975) PQ-tree algorithms. Doctoral thesis, dpt of electrical Engeneering and Computer science. University of California, Berkeley, CA,USA.

2- Eric Boneru and Jonathan Atkin and Bruce Hendeickson (1999) A Spectral Graph Algorithm for Consecutive Ones Property. SIAM Journal Computer 28, 297-310

3- K. Booth and G. Lueker (1976) Testing for the Consecutive Ones Property, Interval Graphs and Graph Planarity Using PQ-tree algorithms. Journal of Computer and System Science 13, 335-379.

4- Mohammad Taghi and Yashar Ganjali (2002) A Note On Consecutive Ones Submatrix Problem . Department of Computer science. University of Waterloo, ON, Canada, information process letter 83,163-166.

5- Davaatseren Baatar. Horst W. Hamacher. Matthias Ehrgot. Gerhard J. Woe (2004) Decomposition of Integer Matrices and Multileaf Collimator Sequencing. **Discrete Applied Mathematics archive 152 ,1-3.**

6- D.R. Davis and G. Kendall (1969) Incidence Matrices and Interval Graphs and Seriation in Archaeology. Pacific journal of Mathematics 28.

7- D.R. Fulkerson and O.A.Gross (1965) Incidence Matrices and Interval Graphs. Pacific journal of Mathematics ,15, 3.

8- Jitender.Deogun and K. Gopalakrishnan.(1999) Consecutive retrieval property-revisited, Information Processing Letters 69,15-20.

9- W.L. Hsu and R. M. McConnell (2003) PC Trees and Circular-Ones Arrangements. **Theoretical Computer Science 99-116. ...**

10- A. Jennings (1966*) A compact storage scheme for the solution of simultaneously equations. Computer Journal 9, 281-285.*

11- Joao Meidanis. Oscar Porto. Guilherme.Telles (1998) On the Consecutive Cnes Property, Discrete Apll .Math 88, 325-354.

12- A.Tucker (1972) A structure theorem for the consecutive ones property. Journal of Combinatorial Theory 12, 153-162.

13- L.T. Kou (1977) Polynomial complete consecutive information retrieval problems. SIAM Journal on Computing 6, 67-75.

14- S.Haddadi (2003) A note on the NP-hardness of the consecutive block minimization problem. International Transactions in Operational Research 9, 775777.

15- M. Habib. R.M. McConnel. C. Paul. L. Viennot (2000) Lex-BSF and Partition Refinement, with Applications to Transitive Orientation, Interval Graph Recognition and Consecutive Ones Testing. Theoretical Computer Science 234, 59-84.

16- D.S. Hochbaum (1997) Approximation Algorithms for NP-Hard Problems (edited by D.S. Hochbaum). PWS Publishing Company, Boston, MA.

17- R.W.Lang, Francis.Lau, and Yingchao Zhar (2003) Consecutive Ones Block Symmetric Matrices. Department of Computer Science and Information Systems

18- J. Edmonds (1965) Paths Trees and Owners. Canadian Journal of Mathematics 17, 449-467.

19- D.S.Greenberg and S.Istrail (1995) Physical Mapping By STS Hybridization Algorithmic Strategies and Challenge of Software Evaluation .Journal of Computational Biology 2, 219-273

20- P. W. Goldberg, M. C Golumbic, H. Kaplan, R. Shamir (2002) Four Strikes against Physical Mapping of DNA. Journal of Computational Biology

21- J. Atkins and M. Middendortf (1996) On Physical Mapping And The Consecutive Ones Property for Sparse Matrices. Discrete Applied Mathematics 71, 23-40

22- Jinsong Tan and Louxin Zhang (2005) Approximation Algorithm for the Consecutives Ones Submatrix Problem On Sparse Matrix. Department of Mathematics. National University of Singapore 2 Science Drive 2, Singapore 117543(BMRC01/1/21/19/140.

23-- W.L.Hsu and C.H.Ma (1991) Fast and Simple Algorithms for Recognizing Chordal Comparability Graphs and Interval Graphs. Lecture Notes in Computer Science 557, 52-60, to appear in SIAM J. Computer.

24- L. Carsten and M. Yannakakis (1994) On the hardness of approximating minimization problems, Journal of the ACM 41,960-981

25- N. Christofides (1976) Worst-case analysis of a new heuristic for the travelling salesman problem, Report 388, Graduate School of Industrial Administration, Carnegie-Mellon University, Pittsburgh, PA, USA,

26- I.Lawler. E.L. Lenstra. J.K. Rinnooy. A.H.G Kan. D.B. Shmoys(2001) The Travelling Salesman Problem. A Guided Tour of Combinatorial Optimization. Wiley and Sons.

27- M.R. Garey and D.S. Johnson (1976) Some simplified NP-complete graph problems. Theoretical Computer Science 1(3), 237-267.

28- S. Istrail and S. Greenberg (1995) Physical mapping by STS hybridization: algorithmic strategies and the challenge of software evaluation. **Journal of Computational Biology 2,** *219-274.*

29- S. Benzer (1959) On the topology of the genetic fine structure PNAS,45,1607- 1620.

30- Burkard (2006) An annotated bibliography of combinatorial optimization problems with fixed cardinality constraints. Discrete Applied Mathematics. Vol 154 , Issue 9,1344 - 1357.

31- Meidanis and E.G. Munuera (1995) A simple linear time algorithm for binary phylogeny. In Proceedings of the XV Int. Conf. of the Chilean Computer Science Society, pages 275-283, Arica, Chile.

32- S. P. Ghosh (1972) File organization: the consecutive retrieval property. Commun. ACM, 15 802-808.

33- D. Gusfield (1997) Algorithms on Strings, Trees and Sequences. Cambridge University Press.

34- Z. Layouni and S. Haddadi (2008) Consecutive Block Minimization is 1.5 Approximable. **Information Processing Letters, v.108 n.3, p.132-135.**

Kolibri 360 GmbH
Ferdinand-Jühlke-Straße 7
99095 Erfurt, Deutschland
produktsicherheit@kolibri360.de

Druck:
CPI Druckdienstleistungen GmbH
im Auftrag der
Kolibri 360 GmbH
Ferdinand-Jühlke-Str. 7
99095 Erfurt